Contents

List of Figures

List of Tables

INTENTIONALLY LEFT BLANK.

1. Introduction

This report is an overview of the past, current, and projected future role of odor in military and security operations. The specific focus of the report is the role of odor in stealth operations, that is operations indented to deceive the enemy by false military force projections and to make the real friendly forces invisible to adversary senses and sensors. In any situation where one entity hunts another, there are two imperatives to survival and success that are equivalent for each entity: the need to avoid or at least mislead detection, and the need to penetrate the equivalent concealing tactics by the other party. This holds as much in the animal predator/prey universe as in the human world of war.

Odor is a property of some gases or chemical substances carried by air that can be perceived by the sense of smell, also called the sense of olfaction, after inhaling (vertebrates) or on contact with antennae (arthropods) by a living organism. Most odors indicate the presence of an organic compound, although some simple chemicals not containing carbon (e.g., ammonia) can also produce an odor. The word "odor"[1] has many synonyms that have positive (aroma, attar, balm, bouquet, fragrance, incense, perfume), neutral (flavor, redolence, scant, smell), and negative (fetor, malodor, mephitis, niff, pong, reek, stink, stench) connotations. Specific odors carry valuable information about environments and/or activities; this information is often critical for a Soldier's situation awareness. Odor may in fact be the first indication of an adversary's actions.

ARL Special Report ARL-SR-242 *Owning the Environment: Stealth Soldier—Research Outline* (May 2012) presented an outline of the visual and auditory research needed in support of future military stealth operations, force projection capabilities, and misdirection and deception activities, and put this research into historical context. Visual and sonic deception and operational discipline (stealth) in war are well-recognized elements of stealth and are equally well represented in the animal world, ranging from sound-muffling feathers in owls to adaptive camouflage in octopi. Odor has not been as much of a concern in human warfare because human noses are rather poor examples of that sense compared to the animal world. For example, dogs have about a 17 times larger and a hundred times more densely innervated reception area (olfactory epithelium) that detects smells, compared to humans (Bear et al., 2006; pp. 265–275). However, the interest in smell in military environments is rapidly growing due to common deployment of domesticated animal companions in the field, e.g., dogs for drug detection at checkpoints. This interest is also due to rapid development of electronic equivalents for both remote and local sensing of chemical traces in the environment. Some of the electronic sensing systems have been designed to detect chemical and biological warfare agents, but recently more capability is being developed in the medical field for physiological status sensing, e.g., detecting

---

[1]The word "odor" has a generally neutral meaning, although in the United States it has a more negative connotation. In this report the term "odor" is used in its neutral meaning.

lung cancer from exhaled breath. These developments, along with a growing realization of the intimate connection of even unconscious perception of odors by humans and their subsequent effects on cognition, raise the need to directly address the topics of odor detection, concealment, misdirection, and possible weaponization. This report first provides background on the sense of smell, covering some roles of odors in life, some aspects of the historical use of odor in human interactions, and the use of odor by animals and plants in the wild. The report then proposes a possible research plan to both prevent surprise as well as create an advantage for the tactical employment of odor. The ultimate goal of this report is to provide an overview and research outline for the sense of smell to facilitate its inclusion in multi-sensory stealth research and technologies as another military "sword and shield."

2. Odors – The Human Experience

The sense of smell (olfaction) is one of the most ancient evolutionary senses and allows organisms with smell receptors to identify food, mates, predators, and changing surrounding conditions. The human sense of smell is mediated by the olfaction sensory cells (epithelial cells) located in the roof of the two nasal cavities of the nose. Each epithelial cell has cilia (hair) with direct contact to the air. Odor molecules act as a chemical stimulus. They bind to receptor proteins extended from the cilia, creating an electrical signal which is then carried to the brain by the olfactory nerve.

Not all chemicals have odor. An odorant, i.e., a chemical that makes an odor, must be volatile or aerosolized (able to float through the air) (e.g., Amoore et al., 1964; Wolfe et al., 2011). All human odorants are non-ionic (in order to be sufficiently volatile) and are usually organic compounds with relative molecular weights of less than 300. The strength of a smell sensation depends on the concentration of odorant molecules in the air. This in turn depends upon factors such as the odor's volatility, environmental wind direction, flow rate, temperature, humidity, and the spatial and temporal nature of the odor source (Cseh et al., 2010). Smell recognition is a reaction to the global mix of all odorous molecules rather than to the concentration or intensity of any single odorant (Axel, 1995; Spengler et al., 2000). It can be said that the sense of smell is a relatively poor analytical sense but an effective classifier that synthesizes the sum of acting stimuli[2]. In general, people can detect a 5–10% change in odor intensity, but odor concentration needs to be increased threefold ($3\times$) above its detection threshold in order for the odor to be identified (e.g., Goldstein, 2007).

The ancientness of the olfactory system in evolutionary history and its direct connection to critical areas in the formation of emotional and autobiographic memory implicate olfaction as a

---

[2]Although most untrained people cannot identify individual components in an odorous complex, specially trained experts, such as perfumers and flavor chemists, can pick out individual chemicals in complex mixes through smell (Herz, 2008).

primal sense governing behavior. Odors are reasonably considered as harbingers of disease, markers of physiological state, and modulators of memory and emotion. The word "malaria," which means "bad air," appears to be derived from the conditioned association between the stench of stagnant water and contracting malarial disease. The association between a smell and illness is powerful, often conditioned in a single trial, and resistant to extinction. In respect to humans, smell is a major factor of situation awareness.

Smell is extremely important to humans, despite the fact that humans are anatomically and genetically not equipped to be powerful smellers. As mentioned previously, humans have fewer olfactory receptors than many other creatures, and although we have about 1000 olfactory receptor genes, >60% of these are nonfunctional (Gilad et al., 2003). It is thought that humans' excellent vision, particularly color vision, might have compensated for evolutionary losses in olfactory ability. Indeed, apes and monkeys with full trichromatic color vision also have a large proportion of nonfunctional olfactory receptor genes, relative to primates with poorer color vision (Gilad et al., 2004). Although humans are not as sensitive to smells as most other animals, they have a surprisingly good sense of smell for their relatively small number of olfactory receptor genes, and their covert and overt reactions to smells are very strong (e.g., Herz, 2008). This seems to be due to humans' relatively large olfaction-related brain region and contribution of higher cognitive processes in smell perception (Shepherd, 2004). Some scents create strong sensual pleasures and are effective aphrodisiacs while others create strong disgust and even fear. The presence of odor enhances the memory of an event. Memories recalled by odors are more vivid and emotional than those recalled by auditory or visual stimulation alone (Herz, 2004). An odor that was associated in memory with previous unpleasant or frustrating activities may even negatively affect future activities if concurrently present (Epple and Herz, 1999). However, some of these views are criticized as based more on individual observations and theoretical considerations rather than on solid empirical evidence (Gilbert, 2008).

The connection with fear is very important for olfactory warfare and also intriguing from a scientific point of view (Krusemark and Li, 2012). It is well documented that most of the substances that have a strong bad odor are not harmful while some subtle and neutral odors may indicate the presence of danger (e.g., GDHR, 2004; NHDES, 2012). Yet, there are many examples in history where the combination of fear with a harmless but strong smell has caused panic or an outbreak of mass hysteria (Bartholomew and Wessely, 2002; Pain, 2001). So, why are humans so suspicious and afraid of a bad smell? Smell is an ancient sensory system in the history of evolution, and its neural wiring has undergone millions of years of selection for enhancing survival (Vokshoor and McGregor, 2012). Olfactory signals travel in the parallel neural pathways from the *olfactory epithelium* to the thalamus and cortex, where the signals are converted into conscious awareness of smell, and to the limbic region of the brain (amygdala), where arriving signals create unconscious emotions (Pain, 2001). As such, an unpleasant odor can degrade human focus, diminish productivity, and increase distaste for a conducted operation. One example of smell-triggered fear relates to the brain wiring of our sense of smell that evolved millions of years ago when the smells of rotten fish (ammonia) and rotten eggs (hydrogen sulfide) indicated real harm to humans. Both these smells are produced by the decay of proteins (putrefaction), and if early humans were not paying attention to these smells, many of them would have died soon from food poisoning. The unpleasant smells might have had the same importance even 100–200 years ago when our abilities to preserve food were still very limited and bad smell was an indication that the food was getting rotten (Turin, 2007). These smells of decay are still important in food safety today, even if modern preservation methods mean we encounter these smells less often. In addition, the strong human memory for smells makes smell an easy flashback signal of previous fear or terror states (Zald and Pardo, 1997; Zald, 2003). With our current focus on potential terrorist activities, a new, unknown smell also becomes suspicious as a potential sign of terrorist gas attack. It may thus be that the presence of strong unpleasant or unknown smells is encoded in our brain as a warning signal that our health is or may be endangered. This may be why *malodorants*, chemical compounds characterized by highly unpleasant stench, act as temporary incapacitants and are used in this capacity by military and security forces.

On the flip side, pleasant odors can also affect human behavior and task performance, for example, by evoking heightened awareness and improved vigilance. It has been reported that subjects perform better in visual target detection when exposed to an arousing scent (peppermint) than when exposed to a relaxing scent (bergamot) (Gould and Martin, 2001).

Individual people tend to prefer certain odors, using them for their own pleasure and to mask other more offensive scents. Some preferences are based on beliefs, others on non-olfactory associations, and still others on past emotions and experiences. Arguably some of these preferences are based on the fact that humans can differentiate individual odors based on kinship, as has been demonstrated in animal models (Zhang, 2011). These odors may arise from shared familial or cultural diet, genome, hygienic practice, or some combination. Historically, military regiments came from geographically confined areas with genomically similar individuals who were kept together to maintain loyalty and cohesion, as shown by many of the names and histories recorded from the U.S. Civil War, such as the 5th New York Volunteer Infantry and the 54th Regiment of the Massachusetts Volunteer Infantry. One could hypothesize that some of the group identity and cohesion was maintained via in-group shared chemical signals, which are now assayable in zeptomole (one sextillionth) quantities. There is evidence that trust and other affiliated behaviors are facilitated by odors (e.g., Bower, 2005). These are governed by oxytocin and are reflected in functional brain imaging studies. Olfactory mechanisms indicating sensitivity to pheromones are evident in neural activities implicating the presence of trust (Liljenquist, 2010), and occur even in social insects (Chapuisat, 2009).

The expectations and beliefs associated with strong unpleasant odors discussed above are just some examples of human preconceptions related to odors (Dalton, 2012). In some cases the judgment of odor is based on non-olfactory sensations and expectations (Kieran, 2010). For example, Brochet (2001) demonstrated that wine poured from a respectable old wine bottle was judged by a majority of wine experts as having a better scent than the same wine poured from a cheap-looking wine bottle. These multi-sensory associations of odors must be remembered in any application of odors for stealth operations and deceptive activities.

3. Odors and Chemical Tactics in Nature

Odor is a critical element of any habitat. Chemical odors abound in nature and can provide animals with important information about their environment (i.e., situation awareness). Odors may indicate the presence of predators or prey, the location of valued resources, the boundaries of home territories, or even the physiological states of other individuals. Odors are also used tactically by plants and animals to attract, to repel, to communicate, and to deceive.

3.1 Finding Food

Predators are drawn to the scents of prey and use these scents to track prey. For example, cats, foxes, and snakes are attracted to mouse scent (Hughes et al., 2012). Predatory fishes and invertebrates also use odor to locate prey (e.g., Atema et al., 1980; Ferner and Weissburg, 2005). The use of smell to attract predators is dramatically illustrated by the use of chum to attract sharks. Attracting these animals is as simple as entering a suitable habitat and then lacing the water with bits of fish and blood. Well-known for their exceptional olfactory abilities, sharks use time delays between their nostrils to localize a smell source (Gardiner and Atema, 2010), which they follow to feeding opportunities.

Parasitic animals also use their sense of smell to find food, in this case, suitable hosts. Mosquitoes are strongly attracted to human foot odor (Knols et al., 1997), and ticks use scent to locate optimal body areas for parasitism (Wanzala et al., 2004). Parasitic vertebrates, such as vampire bats, also use olfaction to locate hosts to feed upon (Bahlman and Kelt, 2007).

The attraction of hungry animals to smells can be exploited to lure animals to particular locations, a tactic well developed by many plants. The strong sweet scent of flowers attracts pollinators like bees and butterflies, and they fertilize fragrant plants in exchange for a sip of nectar. Carnivorous plants also employ scents, including floral-like scents, to lure their insect prey (Jürgens et al., 2009). Orchids attract insects as pollinators, but instead of using food smells they mimic female insect pheromones, attracting the attention of eager males (e.g., Schiestl et al., 2003). Some predatory animals employ this same tactic, emitting the sexual pheromones of their prey as a lure to attract a meal (Gemeno et al., 2000). Caterpillars of *Maculinea* butterflies also mimic the chemical identity signals of other animals in order to obtain food. By smelling like ant larvae, they can fool an ant colony into sheltering and feeding them (Schönrogge et al., 2004; Forbes, 2011).

3.2 Staying Safe

Prey animals recognize the natural scents of predators and respond with defensive behavior (reviewed in Kats and Dill, 1998). Identification of predator scents as dangerous is instinctive or learned in different animals. Naïve mice show fear responses upon smelling cat odors for the first time (Papes et al., 2010). Many fishes quickly learn that a given smell correlates with predator presence and respond with fear upon subsequent exposure (Ferrari et al., 2010). Defensive responses to predator odors can include freezing, avoidance, hiding, or social anti-predator tactics (Apfelbach et al., 2005). Prey can also respond to the odors of what a predator has been eating. Odors of consumed prey are excreted by the predator's digestive system, and these scents trigger anti-predator behavior in conspecific or similarly-sized prey animals (Smith, 1992; Nolte et al., 1994; Wisenden, 2000; Ferrari et al., 2010).

Some potential prey animals use odors to communicate directly with predators, i.e., to advertise that they are dangerous or unpalatable prey and should not be pursued. Some poisonous or unpalatable animals use malodorous secretions, often combined with bright warning colors, to advertise their toxicity to potential predators (Guilford et al., 1987; Moore et al., 1990; Lindstron et al., 2001). Non-poisonous animals can protect themselves from predation by mimicking these warning odors (e.g., Dettner and Lieper, 1994). Odorous chemicals can also be used directly as defense; for example, the two-spotted stink bug protects itself from predators by emitting a volatile liquid from the end of its abdomen (Krall et al., 1999). Plants can use odors to defend themselves as well. Many plants release odorous chemicals when their leaves are damaged by herbivores. These odors attract predators, which then feed on the herbivores and help spare the plant (e.g., Hoballah and Turlings, 2005).

3.3 Chemical Alarm Signals

Many animals release alarm odors when disturbed or startled (e.g., Müller-Schwarze et al., 1984; Parejo et al., 2012) or when injured (Smith, 1992; Wisenden, 2000; Ferrari et al., 2010). Some fishes and amphibians have special chemical packets in their skin that release odor only when damaged. During a predatory attack, these ruptured scent packets release chemical signals into the water. These alarm signals act as warnings to kin, to social group members, and to other nearby prey, enhancing their chances of escape.

3.4 Olfactory Camouflage

Proper use of camouflaging scents can protect animals from predators and rivals. Many animals produce endogenous scents for use as camouflage. Subordinate males in some species will secrete female pheromones as camouflage to avoid being attacked by aggressive larger males (Peschke, 1987). Animals are also known to apply scents from the environment to themselves for camouflage. For example, ground squirrels will chew shed snake skins and rub the snake scent over their fur; this helps protect them from attack by rattlesnakes (Clucas et al., 2008).

3.5 Animal Navigation

Olfaction is important in navigation. Animals use smells to find their way not only to food sources but also to find living or spawning sites and to follow members of their social groups. Many migrating animals find their way back home by memorizing the unique odor profile of their birth location. For example, salmon return to spawn in their home stream, a feat accomplished by memorizing and later matching the unique odor profile of their birthplace (Dittman and Quin, 1996). Animal-generated odors can be used to mark pathways to desired resources. For example, foraging ants lay down scent trails that can be followed by colony-mates (Morgan, 2009).

3.6 Communication and Social Behavior

Olfaction plays a major role in the social lives of group-living animals. For many animals, olfaction may be a more important mode for social communication than vision or hearing.

Social uses of odor are familiar to many people due to their use by companion animals. Dogs are well known to urinate on posts around the neighborhood, while cats will rub their cheek glands and scented saliva onto stationary objects, including their human companions, to mark territory. Rabbits likewise rub their chin gland secretions onto cages and furniture to mark territory and advertise status. Wild animals have similar scent-marking behaviors. Many ungulates have specialized glands near their eyes, cheeks, ankles, or feet for scent-marking shrubs and tall blades of grass. Hyenas, leopards, bears, and other large carnivores will deposit urine and gland-derived scents on trees, rocks, and tall grass. Giant pandas will even do a handstand in order to deposit scents high onto tree trunks (White et al., 2002). Among primates, lemurs have perhaps the most complex olfactory social communication system, producing over 100 different scent compounds from a variety of glands (delBarco-Trillo et al., 2012).

Scent-marking is a means of advertising social status, territoriality, identity, and group membership. Scents left by one animal can be inspected by others in order to determine when a mark was left and which individual left it. Animals can use scent marks and/or direct inspection of bodily odors to recognize individuals (e.g., Brennan and Kendrick, 2006), identify family members and kin relationships (e.g., Mateo, 2003; Mehlis et al., 2008), assess group membership (e.g., Bull et al., 2000; Dapportol et al., 2007), determine territory ownership (e.g., Simons et al., 1994; Zuri et al., 1997; Brashares and Arcese, 1999), identify sex and reproductive states (e.g., Eisenberg and Kleiman, 1972), and ascertain health status (e.g., Zala et al., 2004). This information is helpful for discriminating friend from foe and for determining when to mate, when to hide, or when to fight.

4. Odors and Humans

4.1 Hunting

Most animals are very sensitive to odors and elimination or reduction of human odor is very important to hunters. Books, magazines, and internet pages on hunting devote a lot of attention to smell reduction and, more generally, to methods on how to hide the presence of hunters from their prey. Some of the advice given to hunters includes:

- Do not use soap or other cosmetics.
- Use baby powder as antiperspirant.
- Do not eat spicy food.
- Do not smoke.
- Brush your teeth with baking soda (sodium bicarbonate; odor absorbing chemical).

The general rule is to use hunting clothes made from odor absorbing fibers (e.g., carbon fiber), wash the clothes with unscented detergent, and hang it out to dry. Carbon fibers contain charcoal that absorbs and helps to hide the smell (Anonymous, 2011).

4.2 Law Enforcement

One area of law enforcement where odor detection and identification is an important activity is the Customs and Border Protection (CBP) enforcement. Various prohibited and potentially dangerous substances have a distinct odor and can be recognized by the sense of smell. Since canines (dogs) have a much better-developed sense of smell than humans, they are trained to recognize and report various odors. At border crossings, customs officers/agents are taught to screen passengers, vehicles, and their luggage for narcotics or other prohibited substances with a detection canine. The dogs are trained to "sit" when responding to the odor of a narcotic such as marijuana, cocaine, heroin, methamphetamine, hashish, or ecstasy (CBP, 2012). The dogs are also trained to detect the odor of concealed humans. Inside the country, dogs are used for drug interdiction and explosives detection by highway patrols and for mail sensing (e.g., for hidden substances[3]) by security personnel.

4.3 Disaster Search and Rescue

Canines are also used by Search and Rescue (SAR) Teams in case of natural or human-made disasters and in searching for a missing person. The canines employ their smell detection and recognition skills (e.g., backtracking) in large area searches including snow, desert, pine forest,

---

[3]Some dangerous substances such as anthrax do not have a distinct smell.

and mountain environments. The dogs are also trained in rappelling for helicopter operations to search in remote areas for people or for certain odors (Layton, 2012; Oakes, 2012).

In addition to canines, SAR Teams also use an Electronic Search and Rescue (e-SAR) mass spectroscopy tool that replaces or complements the use of dogs. A sample of the odor is registered in the tool's memory, and sensor data are continuously compared with the stored sample. The tool augments odor sensing with Global Positioning System and wind data for faster operations. People or other living organisms can be detected by respiratory gases, evaporated perspiration and urine (ammonia), or decomposition gases (Tchoukanov, 2012).

4.4 Gas Leaks

Gas leak detection is an important safety procedure to alert people when a dangerous (often combustible) gas has been detected. Before modern electronic sensors were developed, various other procedures had been implemented to warn people about the presence of life threatening gases such as carbon dioxide, carbon monoxide, and methane. For example, coal miners brought canaries down to the tunnels and used them as an early detection system. In the presence of harmful gases, canaries stop singing and eventually die. The other technique was the use of chemically infused paper that turned brown when exposed to the gas. Currently, a variety of electrochemical, infrared, and ultrasonic electronic gas detectors are available that can detect nearly all combustible gases including acetone, acetylene, benzene, butane, ethanol, ethylene oxide, gasoline, hexane, hydrogen, industrial solvents, methane, methanol, paint thinners, propane, natural gas, and naphtha (e.g., Bacharach Leakator).

In the house, a person can detect the presence of a natural gas leak by using a gas leak detector or by smelling a rotten egg odor. In its natural form, natural gas (primarily methane) is odorless, but gas companies add artificial odorants, such as methyl mercaptan and ethyl metrcaptan, to the natural gas so that leaks can be easily detectable by smell.

4.5 Demining and UXO Operations

Unexploded ordinance (UXO) and mines left in the ground after previous wars are of great danger to people re-inhabiting war-tarnished land. There are many various mine and UXO detection devices that have been developed for military and humanitarian use (e.g., metal detectors), but they are still imperfect and result in many false alarms. In addition, some larger and more sophisticated mine detection devices, such as mine clearing vehicles, cannot be used in more rugged terrain. A viable alternative to these devices is the use of animals with a sensitive sense of smell, since the vapors from explosives used in UXOs and mines have odors. The animals most commonly used for detection of vapors from hidden chemicals are dogs and African pouched rats. A human can detect a $1\text{-in-}10^{-4}$ concentration of odors in air, and devices based on gas chromatography can reveal concentrations on the order of $1\text{-in-}10^{-12}$ particles, while dogs can detect concentrations of $1\text{-in}10^{-15}$ and less (Orfici, 2003). The African pouched rat's sense of smell is even more sensitive than a dog's, and the rats are easily trainable. There are an estimated 25 mine clearing organizations in the world using dogs. One of the organizations using African pouched rats is the Belgian Anti-Persoonsmijnen Ontmijnende Product Ontwikkeling (APOPO) operating in Tanzania. The use of Norway rats for smell detection is also being explored (Ferrante, 2012). The advantages of using rats are that they are inexpensive to train and, unlike dogs, are too small to trigger land mines. There were also some attempts to use bees for humanitarian mine detection (Helm, 2005).

One relatively recent technology used for mine detection is the Remote Explosive Scent Tracing (REST) system originally developed in early 1990 in South Africa (Mechem Consultants). The REST system uses multiple cups to collects samples of vapor along segmented (0.5–2.0 km) stretches of road or land. The samples are then brought to a stationary base where they are sniffed by specially-trained dogs and rats to identify (by sitting next to them) which segments have explosives buried in the ground (hit rate 68%; see Jones et al., 2012). The challenge is to train the animals to respond to vapors that are characteristic to explosive material and not to the chemical composition of the land in a specific area.

4.6 Medicine

Popular news stories periodically report on the ability of animals, usually dogs, to use scent to detect medical conditions such as cancer (e.g., Laino, 2012). Released or secreted chemical compounds unique to various cancers can be detected in the breath, the sweat, or the urine of patients. This has sparked attempts to further define this ability and perhaps to develop specific engineered systems that could mimic the animals' diagnostic ability. Gas chromatography has shown some success in this area, but clinical adoption of gas chromatography is limited due to its high costs, both for infrastructure and for actual testing. The use of animals likewise creates a heavy logistical load, due to both the required training as well as upkeep. Electronic noses are being developed for specific diseases such as diabetes, renal disease, and airway inflammation. The electronic noses can distinguish between diseased and healthy breath samples and can even evaluate the efficacy of hemodialysis (treatment for renal failure) (Guo et al., 2010).

Medically important volatiles, such as alkanes and aromatic compounds in lung cancer, can be detected and identified by specifically-built sensors. For example, an electronic nose composed of eight quartz microbalance (QMB) gas sensors coated with different metalloporphyrins can detect lung cancer markers in breath (Di Natale et al., 2003). However, olfactory discrimination does not simply depend on the sensor, but on the processing that underlies the actual transformation from input signals into identified compounds and diagnostic clarity. A dog's nose or other biological system uses a matrix of sensors (receptors) distributed across an olfactory epithelium and uses the pattern of responses across these in order to create the perception of a specific identifiable smell. Attaching an array of different sensors to a neuromorphic computational system may prove to be the most feasible implementation of a system that combines both a wide range of identifications with high levels of sensitivity.

Mohamed et al. (2002) successfully classified the urine of diabetic and non-diabetic patients using electronic nose technology augmented with self-learning artificial neural networks (ANN). Further applications of volatile compound sensing in medicine are only to be expected as sensors and computational sophistication improve.

4.7 Odors and Social Life

People react quickly and emotionally to odors, and they recall smells with greater accuracy than visual or auditory sensations (Sense of Smell Institute, 2012). Memories of faded odors can affect a person's activities and emotions, and specific odors can recall distant (childhood and adolescent) memories (Larsson and Willander, 2009).

In general, females are somewhat more sensitive to odors than are males, and their odor sensitivity and response varies with the menstrual cycle (Doty et al., 1981). Women are also more accurate in odor identification (e.g., Goldstein, 2007; Wolfe et al., 2011). Sensitivity to odors decreases with age and is poorer in smokers than non-smokers (Doty et al., 1984; Goldstein, 2007; Murphy, 1983). Nasal congestion also decreases odor sensitivity. The ability to smell surrounding odors is also conditioned to some degree by the type of surrounding environment. For example, changes in an odor are more distinguishable in cool dry air than in hot humid air (Salthammer and Bahadir, 2009). In addition, human sensitivity to a surrounding odor diminishes with duration of exposure, and people gradually adapt to common odors. For example, a person may cease to notice the odor of cologne or perfume applied a few hours earlier while others still do.

The inability to smell is called anosmia. It is a life-threatening disability. People with anosmia cannot sense the presence of odor-related danger signals such as smoke, toxic chemical smells, or rotten food. According to Hoffman et al. (1998), about 1% of the U.S. population suffers from some form of anosmia. They do not smell odors at all or are insensitive to some types of odors. Anosmia can be a temporary illness due to nasal polyps or inflammation of the nasal mucosa (treated with glucocorticoid sprays such as prednisone), or it can be a permanent impairment caused by head trauma resulting in brain damage or the death of olfactory neural receptors. Olfactory impairment can result from a variety of diseases, such as Alzheimer's and Parkinson's, and may be one of the first indicators that a patient has such a disease (Doty, 2001). Interestingly, olfactory impairment is also associated with psychopathy (Mahmut and Stevenson, 2012). In very rare cases (about 1% of all affected population), people may lack the sense of smell since birth (congenital anosmia) (Karstensen and Tommerup, 2001; Ghadami et al., 2004).

Normally, low level surrounding and unobtrusive odors, such as one's own body odor, are frequently not noticeable until they are paid attention to or their character or level changes. However, even a faint body odor is important for the development of infant-mother attachment, and infants can differentiate their mothers' odors from those of other women (Ferdenzi et al., 2010). Similarly, a mother can discriminate the odor of her own child from that of other infants (Kaitz et al., 1987).

Odors are also very important in attracting people to each other. In courting, people are attracted to odors that are different from their own. Women have a tendency to select partners with body odor indicating the presence of major histocompatibility complex (MHC) alleles different from their own (Wedekind and Füri, 1997; Santos et al., 2005). Such selection ensures that offspring inherit a diversity of genes important for healthy development (Herz, 2002).

Humans are thought to produce a variety of pheromones: chemicals that trigger physiological responses in others when they are absorbed by the recipient's olfactory system. These chemicals may be important in mate attraction and social bonding and may carry information about kinship, social familiarity, fertility, and physical appearance (Grammer et al., 2004). Chemicals secreted by people in fearful conditions (e.g., skydiving) may even act as an alarm pheromone, altering behavior and brain activity in people who inhale these chemicals (Mujica-Parodi et al., 2009; Rubin et al., 2012). There is some debate, however, about how sensitive the human system is to pheromones generally (Doty, 2001).

People are very conscious about their own odor; probably even more than about any other element of their potential reception by others (Li et al., 2007). Thus, while people normally are fairly unaware of their own body odor, they are very sensitive to its changes. These changes are naturally evoked by such causes as consumed food, physical activity, and the emotional state of the person (Chen and Haviliand-Jones, 2000). People living in the same environment, working together, and sharing the same food and customs usually smell similarly and their odor is not noticeable to each other.

One difficulty with sharing information about odors is the richness of vocabulary used to describe odors and a lack of generally accepted classification structure for odors. Since the early days of human history, people have tried to classify odors. Most of these efforts resulted in ordinal scales or hierarchical dendrograms based on qualitative similarity of odors (e.g., Beare, 1906; Cain, 1978; Zwaardemaker, 1895; 1925). One notable classification system of odors was the odor prism proposed by Henning (1916). In this system the odor space had the form of a prism with its six corners defined by six primary odors (flowery, foul, fruity, spicy, burnt, and resinous). All other odors could be described as points in this three-dimensional space. Henning's system did not survive empirical scrutiny but significantly increased interest in smell studies (Wise et al., 2000). More recently, Amoore (1963) proposed that all the existing odors are combinations of seven[4] primary odors listed in table 1. According to Amoore's theory, these seven primary odors differ in the shape or electrical charge of the molecules involved and are the building blocks of all the complex odors existing in nature (Amore et al., 1964). The selection of odors in Amoore's classification was based on specific anosmias, that is, human inability to smell specific odors. The system proposed by Amoore, with its subsequent refining, seems to be getting some support in the physiological literature and, despite its shortcomings (see Rossiter

---

[4]The initial list of seven primary odors has since been expanded to 32 odors.

[1996]; Wise et al., [2000]), it is currently the most-often cited classification system of odors (e.g., Buck and Axel, 1991; Wang et al., 1998).

Table 1. Seven primary odors in nature according to Amoore (1963).

Primary Odor	Example of Odorant	Molecule Form/Charge
Camphor	Mothballs	Sphere
Ether	Dry cleaning fluid	Rod (thin cylinder)
Floral	Roses	Disk with flexible tail
Musk	Aftershave; angelica root oil	Disk
Peppermint	Mint gum	Wedge (prism)
Pungent	Vinegar	Positive charge
Putrid	Rotten eggs	Negative charge

It is important to stress that Amoore's classification is an extension of the shape theory of olfaction (called also the lock-and-key theory) proposed by Moncrieff (1949a; 1949b)[5], stating that both smell receptors and odor molecules come with different shapes and these shapes control the rules for binding odor molecules to the proteins of the olfactory cells. According to this theory and its further extensions, the main way in which smell receptors recognize odors is by the shape of the molecular structure of the chemical compound (Amoore et al., 1964; Goldstein, 2007). This concept has some commonality with an ancient theory of Lucretius[6] (58BC/2010), according to which the sensation of smell results from matching between the size and shape of the nostril's pores and the size and shape of specific odors. An alternative vibration theory of olfaction (called also the swipe card theory) was proposed by Dyson (1938) and expanded by Turin (1996). This theory links smell to vibrational behaviors of molecules in the infrared frequency range. Both theories can be supported by recent physiology findings about the specialized structure of olfactory receptors for which Linda Buck and Richard Axel (Buck and Axel, 1991) received a Nobel Prize (2004) award. According to these findings, perception of odors involves about 1000 distinctly different odor receptors, each of which responds only to a small number of molecular structures. In order for a given odorant to be perceived, its molecules need to bind to several different receptors.

Several other authors proposed limited classifications schemas for odors, but they are far from general and none of them has become a commonly accepted scale (Teixiera et al., 2010). They also do not agree among themselves (Philpott et al., 2008). However, their results can be summarized in several more general observations (Doty, 1975):

---

[5]Moncrieff's shape theory of olfaction is based on the notion of shape-based molecular interactions developed by Pauling (1939), for which the author received a Nobel Prize.

[6]Lucretius (Titus Lucretius Carus), 99BC – 55BC, was a Roman poet and Epicurean philosopher.

1. Two main aspects of odors recognized by humans are intensity (low/high) and valence (pleasant/unpleasant).

2. Increasing the intensity of an odor makes its valence more definite.

3. Most odors and odor names in these schemas are associated with food.

Odor classification schemes are often developed for specific applications. The odor wheel concept lays out a variety of odor classes in one ring and lists specific examples of each odor class in another ring. Odor wheels have been constructed for a variety of odor classification applications, such as for wine, coffee, compost, and wastewater (Burlingame et al. 2004, Rosenfeld et al., 2007, Hammond 2010). Probably the most famous odor wheel is The Wine Aroma Wheel developed for California wines by Ann Noble in 1984 (Noble et al., 1984). The wheel has been modified later by many people and in many ways to be used with other kinds of wines (e.g., in Germany for German wines) but is considered by some experts too limited for current wines (Duman, 2011). Other odor wheels include a beer wheel (Meilgaard et al., 1982), whiskey wheel (Piggot and Jardine, 1979), cider and perry wheel (Williams, 1975), brandy wheel (Jolly and Hattingh, 2001), perfume fragrances wheel (Edwards, 2012, figure 1), compost wheel (Suffer and Rosenfeld, 2007), and fruit juice wheel (Muir et al., 1998).

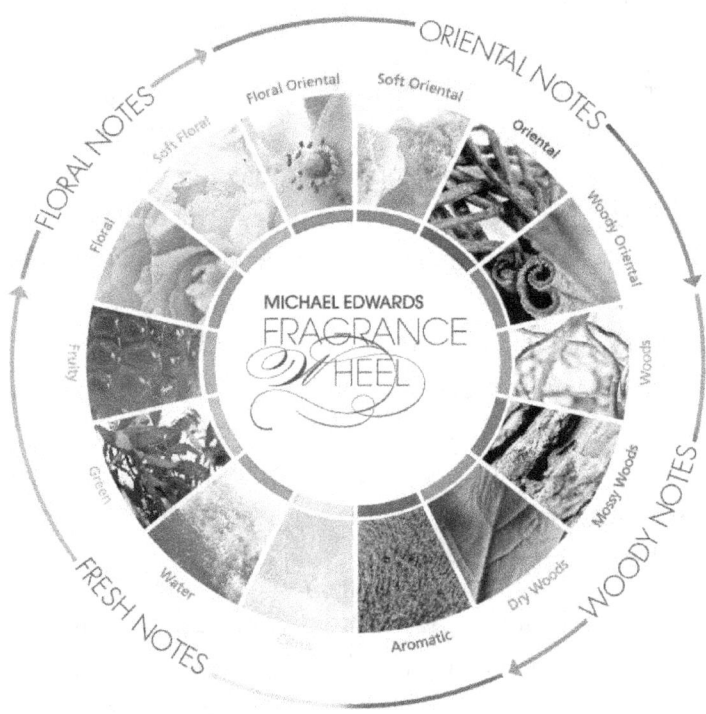

Figure 1. Michael Edwards' perfume fragrance wheel.

4.8 Beauty Industry

The associative power of smell with well-being is often leveraged in the alternative or complementary medicine practices of Aromatherapy[7]. There is also a large industry dedicated to modulating personal odors with deodorants and perfumes.

Humans have been making perfumes for thousands of years (Stoddart 1990), and the practice of producing desirable scents for beauty purposes continues today. The perfume industry is a lucrative one, boasting annual global sales of over $25 billion (Burr 2009). Over 1000 distinct brands of perfume are carried in U.S. department stores (Anonymous 2012). Scented soaps, lotions, shampoos, and deodorants are also widely marketed.

The olfactory beauty industry classifies scents into fragrance families, such as floral, oriental, woody, and fresh. Consumer fragrance preferences are likely affected by both cultural and biological factors, and a synergistic interaction of perfume and natural body odor contributes to perceptions of olfactory beauty (Lenochová et al., 2012). Perfume preferences may function to complement or enhance an individual's natural olfactory signals; people tend to select preferred perfume or cologne scents for themselves in accordance with their genotypes at body odor-related loci (Milinski and Wedekind 2001).

4.9 Entertainment Industry

Smell is such an obvious and intrusive element of life that the entertainment industry has always kept its eye on incorporating smell into games, movies, and music. The first unsuccessful attempts to include smell in the movies were two movie theater systems: AromaRama (invented in1959 by Charles Weiss) and Smell-O-Vision (invented around 1939 by Hans Laube). AromaRama used 31 odors and was used in Carlo Lizzani's movie *Behind the Great Wall* (1959) presented at the Mayfair Theatre in Manhattan, N.Y. A competitive system, Smell-O-Vision, included 30 odors and was used only once in Jack Cardiff's movie *Scent of Mystery* (1960) at the Cinestage Theatre in Chicago. In the former case the odors were distributed through an air conditioning system pushing Freon gas diffusing the odor, and in the latter case the odors were pumped from bottles located on a rotating drum through plastic pipes leading to individual seats in the theater. Both systems used a "scent track" to trigger release of the film's odors. In both cases the odors were weak, the smells persisted longer than was desired, and the molecules were distributed by noisy systems. These olfactory movie systems were financial fiascos.

Another early attempt to incorporate smell into entertainment was Sensorama developed in 1962 (Morton Heilig; U.S. Patent # 3050870) and shown in figure 2. Sensorama was a 4-D arcade game offering five scenarios with 3-D video, sound, and smell. In one of the scenarios the

---

[7]The Sense of Smell Institute introduced the term "Aroma-Chology" to describe the effects of fragrance on behavior. The term "Aroma-Chology" is intended to stress scientific bases of these effects as opposed to the term *Aromatherapy* that stresses anecdotal and healing effects of essential natural oils and herbs (Sense of Smell Institute, 2012).

participant rode a motorcycle through New York while different aromas characteristic of various places in New York were diffused around by surrounding fans (Rheingold, 1992).

Figure 2. Sensorama: 4-D arcade game (1962).

A tribute to the old AromaRama and Smell-O-Vision systems was John Water's comedy *Polyester* (1981) in which the audiences used scratch-and-sniff cards while watching the movie. The cards had 10 numbered spots (1.roses, 2.flatulence, 3.model airplane glue, 4.pizza, 5.gasoline, 6.skunk, 7.natural gas, 8.new car smell, 9.dirty shoes, and 10.air freshener) that the audience scratched and sniffed when the appropriate number flushed at the corner of the screen. This system, called Odorama, solved the problem with hanging odors that was the main problem of the early smell-distributing systems. Scratch-and-sniff cards were later used in the movies *Rugrats Go Wild!* (2003) and *Spy Kids: All the Time in the World* (2011).

A modern version of a smell distributing system for cinemas is the Odorvision system developed by the French company Olf-action and demonstrated at the Lisbon show in 2011. The smells are distributed from dispensers mounted under the ceiling of the cinema. Another similar smell distributing system is SpotScents, developed by Media Information Science Laboratories (MISL) in Japan (Yanagida et al., 2004) for virtual reality (VR) systems. Here, the scent reservoirs are stationary in a ring around the VR platform. The appropriate scent is delivered by air cannon, guided by a video-sensor tracing the position of the participant. The concept of this system is shown in figure 3.

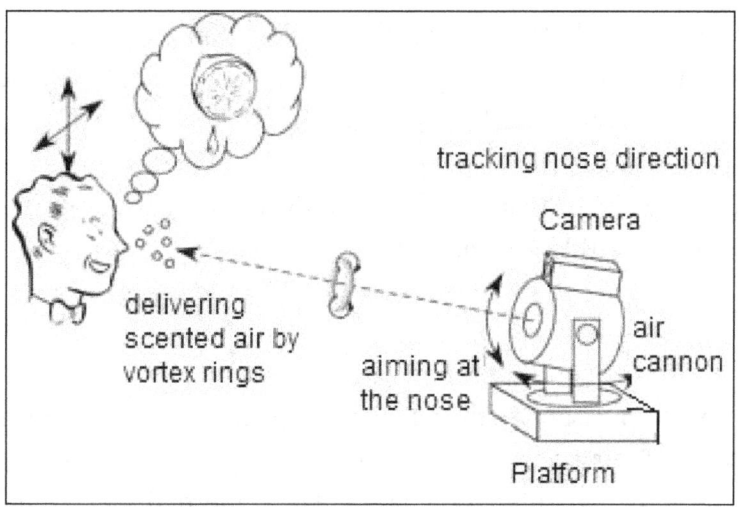

Figure 3. The SpotScents system with air cannons shooting puffs of scents.

Smell is also used in a new version of Terrence Mallick's movie *The New World* (2005) recently distributed in Japan (Fuijwara, 2012), and 20th Century Fox Korea is working on a 4-D version of James Cameron's movie *Avatar* (2009). During the projection of *The New World,* seven smells are emitted by machines placed under seats at the back of the theatre. A floral scent is emitted during love scenes and peppermint and rosemary smells are diffused through the cinema during emotional sequences.

In addition to attempts to add smell to the projection of movies in cinema theaters, several smell-distributing systems were recently (2000−2005) developed to be used with home computers (e.g., iSmell by DigiScents, Pinoke by AromaJet, Smellit by Olf-action), and are currently being developed by companies like Scentcom (www.scentcom.co.il) and AnthroTronics (www.anthrotronics.com). These systems are frequently referred to as digital scent systems. Some of these systems are shown in figure 4.

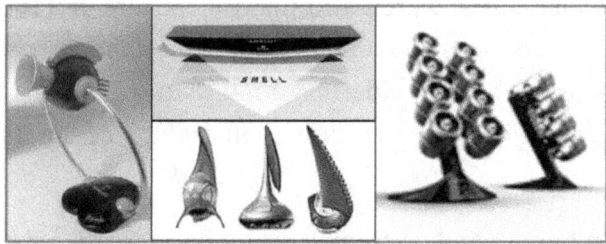

Figure 4. Examples of digital scent systems: Pinoke system
(left), Smellit system by Nuno Teixiera (top), Smellit
system by Olf-action (right), and iSmell system (bottom).

In 2004, France Telecom and Tsuji Wellness demonstrated a device for Internet cafes, which had six cartridges producing different smells. In 2005 two different groups of researchers developed small computer accessories that emitted odors under software control. Currently many companies work on the development of 4-D television (including smell) and commercial 4-D TV sets are expected to be available before the year 2020. For example, researchers from Samsung and the University of California at San Diego (UCSD) developed in 2011 a device that can generate thousands of odors and is small enough to fit on the back of a TV. In this device, an electrical current heats an aqueous solution. The heat builds pressure at specific locations, causing a tiny hole in an elastomer to open, releasing the odor with a dosage controlled by the quantity detector (Kim et al., 2011; Morran, 2011).

4.10 Odor Pollution

Odor is not always a desired or neutral property of an environment or a substance. In many cases, it is a harmless but unpleasant quality disturbing people and making them uncomfortable. Odors that negatively affect quality of life for individuals and communities are air (odor) pollutants, and their emission is regulated by national and international documents. Odors produced by wastewater treatment plants, animal farms, and other facilities can affect human comfort and lead to complaints, decreases in property value, and relocation of residents away from the offending facility. The propagation of odor pollutants through the environment is affected by the type of terrain and vegetation as well as by atmospheric conditions (Cseh et al., 2010). Therefore, proper location of such facilities and proper control of odor emission are major concerns for air quality regulators.

---

5. Odor as Military Agent: Olfactory Warfare

Vision, audition, and smell are the three senses that provide information about the surrounding world without a need for direct physical contact with this world. Visual and sonic stealth and deception are the results of directional, localizable activities. In this respect they differ from the effects of odors, which are much more omnipresent. In addition, the lack of research-based knowledge, norms, and standards in the area of olfactory warfare and the strong dependence of smell propagation on climatic conditions, mean that the commanders in the field must rely more on ingenuity and improvisation when considering smell than when considering sight or sound. Successful stealth and deception operations require consistency of the applied measures across all three sensory inputs.

The review of odor roles in human and nonhuman animal experience, presented in sections 3 and 4, leads to some potential military applications of odorants in olfactory warfare and stealth. The four basic considerations in applying olfactory warfare on the battlefield are consistency with other dimensions of perception, the distance between the odor source and target, the type of environment, and one's own safety (DOD, 1994). Some basic military uses of odorants and odor absorbers are listed in table 2.

Table 2. Applications of olfactory warfare.

Application of Odorant	Comments
Stealth operations (odor level reduction)	A lack of expected smell or no smell at all is the most important protection against adversary ability to detect friendly forces.
Decoy	A smell associated with certain activities can produce a false image of the location of friendly forces, attract attention of the adversary to a certain area in order to neglect another area, or create false alarm.
Deterrent	A strong unpleasant but otherwise harmless odor (fetor, malodor, stench) may prevent people from entering certain protected (no-entry) areas, may stop an adversary from progressing in a specific direction, may create disorganization and panic, and may make some areas secure from close inspection.
Masker/obscurant	A strong neutral or misleading smell can mask smell produced naturally by friendly forces' activities and mask their presence.

The applications listed in table 2 are military applications but they are not unique to military operations. For example, odorants are already used as deterrents in the civilian world. In the U.S., some roadside firs are sprayed with ammonia-based odorants to protect them from Christmas-tree thieves (Pain, 2001).

In recent years, the sense of smell has become an additional sensory warfare aspect in military VR training. Adding olfactory cues to VR simulation together with visual, audio, and haptic cues increases spatial awareness of the user and memory recall of specific environments (Dinh et al., 1999). The Institute for Creative Technologies (ICT) uses the Scent Collar (see figure 5), a scent-emitting device attached to the neck of the participant, to dispense various odors depending on the ongoing training situation in the VR. The current version of Scent Collar[8] has four containers that can be fully open, partially open, or closed depending on the state of the action in the VR simulation (Herz, 2008). Each container contains a fragrance-filled wick and two ports that can be remotely (via Bluetooth) opened or closed for a certain period of time from the VR environment depending on the script. For example, the smell of cigarette smoke in the VR scenario may be an indication that the abandoned-looking building may be in fact occupied.

---

[8]The original version of Scent Collar was developed and patented in 2002. The version had two containers. The current version with four containers was developed in 2005.

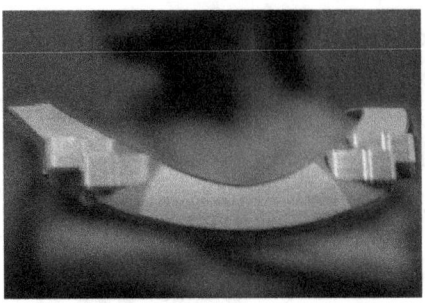

Figure 5. The Scent Collar developed jointly
by ICT and AnthroTronix.

One problem with using olfactory warfare (odorants) is potential noncompliance with the Chemical Weapon Convention and Biological and Toxin Weapon Convention. Odorants can be misjudged by affected people as the presence of biological or chemical weapons. Such a belief may create panic and harm civilian populations. Similarly, people frustrated with smell that they cannot remove may develop both physical and mental health problems (Laumbach et al., 2011).

6. History of Odor in Military Operations

Ancient groups such as Australian Aborigines or American Indians were trackers and stalkers and approached their prey by running from where they had a natural cover or by "freezing" and crawling in the open terrain (e.g., Crystal, 2012). They approached their prey from a downwind direction and frequently painted their faces or their whole body with charcoal, ochre, or mud to disguise their smell (e.g., Rosture, 2012). The pygmies of the Congo rolled around and got covered in fresh elephant dung when they were hunting elephants (Denis, 1963).

Wars offer many examples of using odor to add credibility to decoys or deceiving operations. For example, during the American Civil War, Union general William S. Rosecrans used the odor of burning scraps of wood and the sound of empty barrels to imitate the action of boat-building downstream on the Tennessee River in the historic Battle of Chattanooga. While Confederate forces expected a Union river crossing assault at this location, the Union army crossed the river upstream through a stealthily built pontoon bridge (Foote, 1992).

During the Battle at El Alamein in World War II, British forces used the odor of cordite and diesel to simulate an amphibious assault from wooden rafts deployed out to sea. Simulated odor was also used in deceiving Operation Titanic during the Allied Forces invasion of Normandy (D-Day). A dummy parachute assault including sound sources to reproduce the sounds of battle and canisters of chemicals to reproduce the smell of combat was staged east of La Havre to delay German troops from reaching the real drop zones (Haswell, 1979; Haswell, 1985; Latimier, 2003).

Covering the scents of gunfire or other military activities can be a valuable means of deceiving the enemy. Iran has recently announced the development of a "Deceit Perfume" that attempts to conceal the odor of gunpowder with natural weather-related scents (Beckhusen, 2012).

During the Vietnam War, the U.S. developed and fielded devices called "people sniffers" (XM-2 manpack and XM-3 airborne system), which purpose was to detect Viet Cong and North Vietnamese Army soldiers hiding in a jungle (Anonymous, 1967). The devices detected the ammonia that is found in excreted urine and sweat, but they were not very successful (Dunnigan and Nofi, 2000). The 24-lb XM-2 device (E63 Manpack Personnel Detector) was quickly abandoned since it was primarily detecting the smell of the user rather than of the hiding enemy. The device also had a distinct acoustic signature that was easily detectable (Kirby, 2007). The XM-3 was a little bit more successful since it was also able to detect smoke caused by small arms fire. However, the effectiveness of its ammonia sensor was easily compromised by decoys: Vietnamese soldiers made buckets of mud with traces of urine and hung these buckets in trees or placed them under grass cover.

One of the non-lethal weapons used by military and security forces are stink bombs; devices designed to create an unpleasant smell forcing people to leave an area or protecting off-limits areas against being entered. Such devices range from simple prank devices to military and riot control chemical agents. Stink bombs have existed since WWII and they are getting increased attention in recent years. Some chemicals used in stink bombs include ammonium sulfide, hydrogen sulfide, carboxylic acids, and mercaptans. In 1966 the U.S. Defense Advanced Research Projects Agency (DARPA) and the Battelle Memorial Institute initiated a joint project to develop culturally specific stink bombs, which would affect Vietnamese guerillas, leaving the U.S. troops unaffected. The project was abandoned due to technical barriers.

In 2001 the U.S. announced the development of the ultimate stink bomb aimed at driving away hostile forces by a stench so foul that it results not only in disgust or aversion but also fear. The odorant used in the bomb has been developed by a team of researchers led by Dr. Pamela Dalton at the Monell Chemical Senses Center in Philadelphia and is a mixture of two agents: the U.S. Government Standard Bathroom Malodor (a mixture of eight chemicals with a stench similar to human feces but much stronger) and the Who-Me?, a sulphur-based odorant that smells like rotting carcasses (Mihm, 2002; Pain, 2001; Trivedi, 2002). The latter had been originally developed by the U.S. Office of Strategic Services during WWII to help the French Resistance embarrass German officers by making them smell like rotten meat[9] and limit their appearance in public (Pain, 2001). The odorants and their mixtures have been tested on volunteers of different ethnic origins to make sure that they were universally effective. Both these mixtures are listed in the Guinness Book of Records as the two smelliest substances available on the Earth. Similar preparations are the Smell of Death, a substance found in Israeli Skunk bombs and sprays, which

---

[9]Unfortunately, this substance was so volatile that it could not be confined to specific targets and contaminated everything in the area.

are used for crowd control (McManners, 2004). The odor of Skunk sprays is so persistent that it stays with the skin of the affected person for several days before it can be scrubbed away and can linger on clothes for up to 5 years (Haaretz Service, 2004; Davies, 2008). This persistence has an additional advantage (or drawback – depending on the point of view): the Skunk stench not only disperses a crowd or prevents an adversary from entering an affected area, but it also makes it possible, even several days later, to identify a person who was present at that particular place (Davies, 2008).

Stink bombs can be used in the form of mortar shells, chemical mines, and small projectiles. The odorants are enclosed in small pellets (microcapsules) coated with protective shells (e.g., Durant et al., 2000). The odor is released when pellet is crushed, e.g., by applying physical pressure such as a footfall, or in response to environmental changes such as an increase in temperature.

Harnessing information from natural odors in the environment, such as the individually distinctive scents of different people, can also be quite useful for military and security forces. Current devices such as the E-Nose developed with DARPA can discriminate individual people based on their unique smells, using up-close samples. The Army's Identification Based on Individual Scent (IBIS) program is working to improve this technology to be able to identify persons of interest from afar by using their odor (Dillow, 2010).

7. Odor and Stealth

As discussed in the ARL Special Report ARL-SR-242 Owning the Environment: Stealth Soldier—Research Outline (May 2012), the goals of stealth operations are "to develop novel means to minimize detection of intended activities through sensory diversion and by presenting false information to the enemy about the surrounding environment" (p. 8). Such means include means for hiding our presence and for false force projection to divert enemy attention from our action or to mislead the enemy regarding our intention or strength.

Field Manual FM 33-1-1 (1994) on psychological operations lists four factors that must be considered in using odors in psychological operations (section 5 and appendix A):

- Consistency, that is, compatibility with other measures

- Distance, that is, proximity of the target

- Environment, that is, metrological factors such as wind

- OPSEC (operations security), that is, that activity odors should be masked or eliminated.

Although all these factors are considered very important, consistency is the most important of them. Consistency stresses that any form of deception must affect all senses of the adversary if it is to be believable and effective. The ARL-SR-242 report focused on visual and auditory means of stealth and deception while merely acknowledging the need for the olfactory component to be addressed. The ARL-SR-242 approach to olfaction was limited mainly due to the lack of knowledge about olfactory processes at the cognitive level and because the importance of smell on the battlefield is known primarily through scarce, mostly anecdotal, reports. The role of smell in decision-making processes on the battlefield is largely unknown.

To bridge this knowledge gap, some novel sensory and cognitive research in odor perception is needed. This research needs to be supplemented by the development of reliable means to absorb, emit, and control odors. Both these efforts need to focus on both individual Soldiers and the dismounted squads that are the basic action formations in the current U.S. Army strategic doctrine. The initial focus of military olfactory research must be on basic science and situation awareness, which will further the ultimate goal of stealth applications. Right now, there is much to be learned, for example, about how humans react to odors in the context of concurrent visual and auditory stimuli and about odor perception under stress.

8. Olfactory Research: Outline

Most of the olfactory research conducted to date has been based on individual observations and introspection. More systematic research has been developed in the areas of quality (timbre, color) of smell for beauty aids (e.g., perfumes) and beverages (e.g., wine, beer, brandy). Some research has also been done in determining the spread of olfactory pollutants (stench) emitted by landfills and swine farms. In such research studies, the concentration of odor molecules in air is measured physically (olfactometer) and assessed perceptually by trained panels of odor assessors. Standardized practices recommended in assessment of smell by odor panels can be found in publications by the American Society of Testing and Materials (ASTM) (1991; 1999). Recommended permissible threshold quantities of odor pollutants (in reference to n-butanol) have been specified in European Normalization Standard EN 13725 (CEN, 2001). The training of sensory assessment groups, which may apply to odor assessors as well, is described in ASTM document STP 758 (ASTM, 1981). However, despite some extensive work done in the beauty industry and in support of environmental health agencies, not much research has been done in the area of typical environmental odors and especially in the areas of odor discrimination and recognition. Odor source localization by humans is also under-studied[10].

Military olfactory research should address all potential applications of olfactory warfare listed in table 2: odor level reduction (odor elimination, i.e., for OPSEC purposes), odor as a deterrent, odor as a decoy, and odor as a masker or obscurant. The relative weight attached to each of these

---

[10]It is known that odor sources can be well localized by several species such as moths (Bellanger and Willis, 1996), rats (Bhalla and Bower ,1997), sharks (Gardiner and Atema, 2010), and lobsters (Atema, 1996).

domains should reflect long-term goals of the U.S. Army strategic doctrine and short-term needs for Soldier safety and lethality.

The initial goal of the U.S. Army Research Laboratory's (ARL's) olfactory research is to establish a sufficient foundation for such research in terms of research personnel, basic facilities, and organizational support. An important consideration in planning such research is that it does not duplicate academic olfactory-only research but is conducted from its outset in a multisensory context and with the goal of future field application to stressful military situations and covert operations.

Despite growing interest in the sense of smell both in academic institutions and by industry, there is still a general lack of understanding of the olfactory sense at both psychophysical and psychosocial levels. In addition, both odor-sensing and odor-generating integration mechanisms are not well understood. Therefore, more basic research is needed to understand the sensory, cognitive, and psychosocial process in response to various odors, especially in stressful time-limited situations.

Olfactory memory is robust and resistant to forgetting. How can such memory be used to enhance the human's ability to recall critical lessons learned and to re-visit past activities? The unique smell of an odor results from the activation of a number of olfactory receptors in the olfactory epithelium. Each odor activates a different combination of receptors and potentially increases or decreases the output level from sensory neurons. Two different odors may compete for recognition trying to activate and deactivate the same neurons at the same time. These processes are not well understood and they are critical for understanding olfactory masking phenomena. There is also very limited knowledge about which odors are good maskers of specific smells and about the common interpretation (meaning) of odor combinations by exposed populations. Studies of odor-on-odor masking focused on military relevant odors are yet to be conducted.

For military purposes, an odor that dominates the environment is seldom useful except for being a deterrent (see table 2). A strategic odor needs to blend with natural environmental odors to make the exposed population believe that the odor's release was not intentional. Very little, if any, research has been done on the intensity of specific odors in complex olfactory environments, and it is not yet known how to emit tactical odors that are detectable and recognizable but not suspicious due to their intensity.

While the need for basic research in the area of olfaction is clearly evident, some forms of applied research should also be considered. For example, Moore (2011) found that adding zinc metal particles in picomolar concentration to the air strongly enhances the smell of odors related to explosives. An ARL research plan should consider replication of this finding and explore the possibility of using specially designed zinc particle sprays to enhance explosive detection by both humans and animals. Could such enhancement be useful in mine detection such as securing a safe pathway through an old mine field or in humanitarian demining?

There is also evidence that trust and other affiliated behaviors are affected by common olfactory pheromones. Similarly, unknown odors create suspicion and resistance to cooperation. One can hypothesize that unit cohesion can be enhanced by in-group shared chemical signals. An important research question is how these chemical identities (odor-prints) are formed. What kind of olfactory chemical signals should be considered for an ad-hoc formed action unit to facilitate its cohesion?

The research needs and goals described above have numerous barriers to overcome and different levels of complexity. They can be generally divided into near-, mid-, and long-term goals in order to ensure the most effective and promising execution of this olfactory research plan.

Near-term Goals:

- Establish a material base for olfactory research at ARL's Human Research and Engineering Directorate.

- Conduct an extensive literature review of olfactory research and odor manufacturing.

- Develop a testing methodology for investigation of olfactory sensitivity to odor-related environmental changes.

- Identify salient olfactory characteristics of selected environments suitable for odor manipulation.

- Determine human detection and recognition thresholds for selected military-relevant odors in the presence of well-controlled levels of background odors.

- Measure existing odors and add odor specific data fields to ongoing psychological and decision science projects to develop an in-house database of possible baseline odor effects.

Mid-term Goals:

- Identify a group of odors that need to be masked.

- Investigate propagation properties of selected olfactory maskers.

- Develop a range of concentrations of olfactory maskers that can be used in selected environments under specific operational conditions.

- Develop and field-validate mission-relevant odor manipulation techniques of the surrounding environment.

- Determine the basic properties and operational effectiveness of selected odorants as olfactory decoys.

- Develop means to alter existing odors to make them mask additional odors while preventing the new odors from attracting the attention of the observer (adversary).

Long-term Goals:

- Create an annotated database of olfactory deception techniques and related technologies that Soldiers are currently using to protect and hide their activities.

- Investigate cognitive effects of specific odors in selected operational situations and environments.

- Develop technical means to enable stealth manipulation of odors in the surrounding environment.

- Build prototypes of environmental odor manipulators.

- Develop and field test selected olfactory decoys.

- Develop intelligent odor-dispensing techniques and means (tethered and remotely controlled) responding to metrological conditions and concentrations of other odors in the environment.

- Develop an operational guide and technical means (active smell reduction – ASR) for inhibition of personal (self-) odors or odors created during certain activities.

- Develop controlling means for synchronizing auditory, visual, and olfactory deception to create a multimodal deception strategy.

- Determine the olfactory factors affecting human social behaviors in conducting joint operations.

As discussed previously, the focus of the short-term and mid-term goals should be on general (basic and applied) olfactory research that is still lacking. However, the long-term goals should aim to create specific guidance for human behaviors and technical means supporting Soldiers' olfactory stealth.

9. References

Amoore, J. E. Stereochemical Theory of Olfaction. *Nature* **1963**, *198* (4877), 271–272.

Amoore, J. E.; Johnston, J. W. Jr.; Rubin, M. The Stereochemical Theory of Odor. *Scientific American* **1964**, *210* (2), 42–49.

Anonymous. Applied Science: Sniffing Out the Enemy. *TIME Magazine* **1967**, *89* (23).

Anonymous. Archery and Hunting: Where, How and Why to Use Scent Control Clothing. Available: [http://www.essortment.com/archery-hunting-use-scent-control-clothing-44716.html]. Confirmed on June 4, 2012.

Anonymous. Perfume Industry Statistics. *Statistics Brain.* Available: [http://statisticbrain.com/perfume-industry-statistics]. Confirmed on August 2, 2012.

AnthroTronics Home Page. http://www.anthrotronics.com (accessed Dec 2012).

Apfelbach, R.; Blanchard, C. D.; Blanchard, R. J.; Hayes, R. A.; McGregor, I. S. The Effects of Predator Odors in Mammalian Prey Species: A Review of Field and Laboratory Studies. *Neuroscience and Biobehavioral Reviews* **2005**, *29*, 1123–1144.

ASTM. Guidelines for the Selection and Training of Sensory Panel Members STP 758. West Conshohocken, PA: American Society for Testing and Materials (ASTM), 1981.

ASTM. Standard Practice for Determination of Odor and Taste Thresholds by a Forced-Choice Ascending Concentration Series Method of Limits E679-91. Philadelphia, PA: American Society for Testing and Materials (ASTM), 1991.

ASTM. Standard Practice for Suprathreshold Odor Intensity Measurements E-544-99. Philadelphia, PA: American Society for Testing and Materials (ASTM), 1999.

Atema, J. Eddy Chemotaxis and Odor Landscapes: Exploration of Nature with Animal Sensors. *Biological Bulletin* **1996**, *191*, 129–138.

Atema, J.; Holland, K.; Ikehara, W. Olfactory Responses of Yellowfin Tuna (*Thunnus albacares*) to Prey Odors: Chemical Search Image. *Journal of Chemical Ecology* **1980**, *6* (2), 457–465.

Axel, R. The Molecular Logic of Smell. *Scientific American* **1995**, *273* (4), 154–159.

Bahlman, J. W.; Kelt, D. A. Use of Olfaction during Prey Location by the Common Vampire Bat (*Desmodus rotundus*). *Biotropica* **2007**, *39* (1), 147–149.

Bartholomew, R. E.; Wessely, S. Protean Nature of Mass Sociogenic Illness: From Possessed Nuns to Chemical and Biological Terrorism Fears. *British Journal of Psychiatry* **2002**, *180*, 300–306.

Bear, M. F.; Connors, B. W.; Paradiso, M. A. *Neuroscience: Exploring the Brain* (3rd ed.); Baltimore, MD: Lippincott, Williams, and Wilkins, 2006.

Beare, J. I. *Greek Theories of Elementary Cognition from Alcmaeon to Aristotle*; Oxford, UK: Clarendon Press, 1906.

Beckhusen, R. Perfume of War: Iran Makes Musk to Conceal Troops. *Wired*, Sept. 4, 2012. Available: [http://www.wired.com/dangerroom/2012/09/perfume/]. Confirmed on Sept. 5, 2012.

Bellanger, J. H.; Willis, M. A. Adaptive Control of Odor Guided Locomotion: Behavioral Flexibility as an Antidote to Environmental Unpredictability. *Adaptive Behavior* **1996**, *4*, 217–253.

Bhalla, U.; Bower, J. M. Multi-day Recording from Olfactory Bulb Neurons in Awake Freely Moving Rats: Spatial and Temporally Organized Variability in Odorant Response Properties. *Journal of Computational Neuroscience* **1997**, *4*, 221–256.

Bower, B. Investing on a Whiff: Chemical Spray Shows Power as Trust Booster. *Science News* **2005**, *167*, 356–357.

Brashares, J. S.; Arcese, P. Scent Marking in a Territorial African Antelope: I. The Maintenance of Borders Between Male Oribi. *Animal Behaviour* **1999**, *57* (1), 1–10.

Brennan, P. A.; Kendrick, K. M. Mammalian Social Odours: Attraction and Individual Recognition. *Philosophical Transactions of the Royal Society of London Series B* **2006**, *361* (1476), 2061–2078.

Brochet, F. Chemical Object Representation in the Field of Consciousness. Unpublished manuscript. Prix Coup de Coeur - General Oenology Laboratory. Talence Cedex (France); Académie Amorin, 2001.

Buck, L.; Axel, R. A Novel Multigene Family May Encode Odorant Receptors: A Molecular Basis of Odor Recognition. *Cell* **1991**, *65* (1), 175–187.

Bull, C. M.; Griffin, C. L.; Lanham, E. J.; Johnston, G. R. Recognition of Pheromones from Group Members in a Gregarious Lizard, *Egernia stokesii. Journal of Herpetology* **2000**, *34* (1), 92-99.

Burlingame, G. A.; Suffet, I. H.; Khiari, D.; Bruchet, A. L. Development of an Odor Wheel Classification Scheme for Wastewater. *Water Science Technology* **2004**, *49* (9), 201-209

Burr, C. Perfumers Breathe in Sales Data, and Strategize. *New York Times,* June 19, 2009. Available: [http://www.nytimes.com/2009/06/20/business/20perfume.html]. Confirmed on August 2, 2012.

CBP. CBP canine disciplines. [http://cbp.gov/xp/cgov/border_security/canine/disciplines_2.xml]. Confirmed on July 22, 2012.

Cain, W. S. History of Research on Smell. In: F.C. Carterette and M.P. Friedman (eds.), *Handbook on Perception: Tasting and Smelling*, Vol. VI, pp. 197–229. New York, NY: Academic Press, 1978.

CEN. Determination of Odour Concentration by Dynamic Olfactometry. European Normalization Standard EN 13725. Geneva, Switzerland: European Committee for Standardisation (CEN), 2001.

Chapuisat, M. Social Evolution: The Smell of Cheating. *Current Biology* **2009**, *19* (5), 196–198.

Chen, D.; Haviland-Jones, J. Human Olfactory Communication of Emotion. *Perceptual and Motor Skills* **2000**, *91* (3 Pt 1), 771–781.

Clucas, B.; Owings, D. H.; Rowe, M. P. Donning Your Enemy's Cloak: Ground Squirrels Exploit Rattlesnake Scent to Reduce Predation Risk. *Proceedings of the Royal Society of London Series B* **2008**, *275*, 847–852.

Crystal, E. Australian Aborigines - Indigenous Australians. Available: [http://www.crystalinks.com/aboriginals.html]. Confirmed on June 4, 2012.

Cseh, M.; Narai, K. F.; Barcs, E.; Szepesi, D. B.; Szepesi, D. J.; Dicke, J. L. Odor Setback Distance Calculations Around Animal Farms and Solid Waste Landfills. *Időjárás* [Quarterly Journal of the Hungarian Meteorological Service] **2010**, *114* (4), 303–318.

Dalton, P. There's Something in the Air. In: G. M. Zucco, R. S. Herz, and B. Schaal (eds.), *Olfactory Cognition: From Perception and Memory to Environmental Odours and Neuroscience*, pp. 23–38. Amsterdam, The Netherlands: John Benjamins Publishing Company, 2012.

Dapporto, L.; Santini, A.; Dani, F. R.; Turillazzi, S. Workers of a *Polistes* Paper Wasp Detect the Presence of Their Queen by Chemical Cues. *Chemical Senses* **2007**, *32* (8), 795–802.

Davies, W. New Israeli Weapon Kicks Up Stink. BBC News Report (02 October 2008). Available: [http://news.bbc.co.uk/2/hi/middle_east/7646894.stm]. Confirmed on June 28, 2012.

Dyson, M. The Scientific Bases of Odor. *Journal of the Society of Chemical Industry* **1938**, *57* (28), 647–651.

delBarco-Trillo, J.; Sacha, C. R.; Dubay, G. R.; Drea, C. M. *Eulemur*, Me Lemur: The Evolution of Scent-signal Complexity in a Primate Clade. *Philosophical Transactions of the Royal Society of London Series B* **2012**, *367* (1597), 1909–1922.

Denis, A. *On Safari. The Story of My Life*; London, UK: Collins Clear-Type Press, 1963.

Dettner, K.; Lieper, C. Chemical Mimicry and Camouflage. *Annual Review of Entomology* **1994**, 39, 129-154.

Di Natale, C., Macagnano, A., Martinelli, E., Paolesse, R., D'Arcangelo, G., Roscioni, C., Finazzi-Agrò, A., and D'Amico, A. Lung Cancer Identification by the Analysis of Breath by Means of an Array of Non-selective Gas Sensors. *Biosensors and Bioelectronics* **2003**, *18* (10), 1209–1218.

Dillow, C. The Army Wants Olfactory Sensors that Can Smell Potential Perps at a Distance. Available: [http://www.popsci.com/technology/article/2010-04/army-wants-smelling-sensors-can-id-potential-perps-afar]. Confirmed on August 16, 2012.

Dinh, H. Q.; Walker, N.; Song, C.; Kobyashi, A.; Hodges, L. Evaluating the Importance of Multi-sensory Input on Memory and the Sense of Presence in Virtual Environments. *Proceedings of the 1999 IEEE Virtual Reality Conference*, pp. 222–228. Houston, TX: IEEE, 1999.

Dittman, A. H.; Quin, T. P. Homing in Pacific Salmon: Mechanisms and Ecological Basis. *Journal of Experimental Biology* **1996**, *199*, 83–91.

DOD Psychological Operations Techniques and Procedures. U.S. Army Field Manual FM 33-1-1. Washington, DC: DOD, 1994.

Doty, R. L. An Examination of Relationships Between Pleasantness, Intensity, and Concentration of 10 Odorous Stimuli. *Perception and Psychophysics* **1975**, *17*, 492–496.

Doty, R. L. Olfaction. *Annual Review of Psychology* **2001**, *52*, 423–452.

Doty, R. L.; Shaman, P.; Applebaum, S. L.; Giberson, R.; Sikorski, L.; Rosenberg, L. Smell Identification Ability: Changes with Age. *Science* **1984**, 226, 1441–1443.

Doty, R. L.; Synder, P. J.; Huggins G. R.; Lowry, L. D. Endocrine, Cardiovascular, and Psychological Correlates of Olfactory Sensitivity Changes During the Human Menstrual Cycle. *Journal of Comparative and Physiological Psychology* **1981**, 95, 45–60.

Duman, D. J. Rethinking the Wine Aroma Wheel. *Huff Post Food*. Available: [http://www.huffingtonpost.com/david-j-duman/rethinking-the-wine-aroma-wheel_b_838700.html], 2011. Confirmed on August 17, 2012.

Dunnigan, J. F.; Nofi, A. A. *Dirty Little Secrets of the Vietnam War: Military Information You're Not Supposed to Know*; New York, NY: Macmillan Books, 2000.

Durant, Y.; Thiam, M.; Petcu, C.; Vashista, N. Developing Microcapsules for NLW Applications. Presentation at the Non-Lethal Technology and Academic Research Symposium (NTAR-2000). Portsmouth, NH, 15–17 November 2000. Available: [http://www.unh.edu]. Confirmed on July 20, 2012.

Edwards, M. *Fragrances of the world: 2012 Edition* (28th ed.); Sydney (Australia), 2012.

Eisenberg, J. F.; Kleiman, D. G. Olfactory Communication in Mammals. *Annual Review of Ecology and Systematics* **1972**, *3*, 1–32.

Epple, G.; Herz, R. S. Ambient Odors Associated to Failure Influence Cognitive Performance in Children. *Developmental Psychobiology* **1999**, *35* (2), 103–107.

Ferdenzi, C.; Schaal, B.; Roberts, S. C. Family Scents: Developmental Changes in the Perception of Kin Body Odor? *Journal of Chemical Ecology* **2010**, *36* (8), 847–854.

Ferner, M. C.; Weissburg, M. J. Slow-moving Predatory Gastropods Track Prey Odors in Fast and Turbulent Flow. *Journal of Experimental Biology* **2005**, *208*, 809–819.

Ferrante, J. Rats Go High Tech to Root Out Land Mines. Bucknell University. Available: [http://bucknell.edu/x77893.xml]. Confirmed on Aug. 9, 2012.

Ferrari, M.C.O.; Wisenden, B. D.; Chivers, D. P. Chemical Ecology of Predator-prey Interactions in Aquatic Ecosystems: A Review and Prospectus. *Canadian Journal of Zoology* **2010**, *88*, 698–724.

FM 33-1-1. *Psychological Operations: Techniques and Procedures*. Field manual: U.S. Department of the Army. Washington, DC: DOD, 1994.

Foote, S. *The Civil War: A Narrative* (3 vols.); London, UK: Pimlico Books (Random House), 1992.

Forbes, M. Masters of Disguise: Animal Mimics Fool Their Foes. *Scientific American* **2011**, *304* (6).

Fujiwara, C. Wake Up and Smell the New World. Available: [http://www.filmlinc.com/index.php/film-comment-2012/article/wake-up-and-smell-the-new-world]. Confirmed on August 14, 2012.

Gardiner, J. M.; Atema, J. The Function of Bilateral Odor Arrival Time Differences in Olfactory Orientation of Sharks. *Current Biology* **2010**, *20* (13), 1187–1191.

GDHR. *Health Effect of Odors*. Information Flyer DPH04/239HW. Atlanta, GA: Georgia Department of Human Resources, 2004.

Gemeno, C.; Yeargan, K. V.; Haynes, K. F. Aggressive Chemical Mimicry by the Bolas Spider *Mastophora hutchinsoni*: Identification and Quantification of a Major Prey's Sex Pheromone Components in the Spider's Volatile Emissions. *Journal of Chemical Ecology* **2000**, *26*, 1235–1243.

Ghadami, M.; Morovvati, S.; Majidzadeh-A, K.; Damavandi, E.; Nishimura, G.; Kinoshita, A.; Pasalar, P.; Komatsu,K.; Najafi, M. T.; Niikawa, N.; Yoshiura, K. Isolated Congenital Anosmia Locus Maps to 18p11.23-q12.2. *Journal of Medical Genetics* **2004**, *41*, 299–303.

Gilbert, A. *What the Nose Knows: The Science of Scent in Everyday Life*; New York, NY: Crown Publishers, 2008.

Gilad, Y.; Paabo, S.; Lancet, D. Human Specific Loss of Olfactory Receptor Genes. *Proceedings of the National Academy of Sciences* **2003**, *100* (6), 3324–3327.

Gilad, Y.; Wiebe, V.; Przeworski, M.; Lancet, D.; Paabo, S. Loss of Olfactory Receptor Genes Coincides with the Acquisition of Full Trichromatic Vision in Primates. *PLOS Biology* **2004**, *2* (1), e5.

Goldstein, E. B. *Sensation and Perception* (7th ed.); New York, NY: Thomson-Wadsworth, 2007.

Gould, G.; Martin, G.N. A Good Odour to Breathe? The Effect of Pleasant Ambient Odour on Human Visual Vigilance. *Applied Cognitive Psychology* **2001**, *15* (2), 225–232.

Grammer, K.; Fink, B.; Neave, N. Human Pheromones and Sexual Attraction. *European Journal of Obstetrics and Gynecology and Reproductive Biology* **2004**, 118 (2), 135–142.

Guilford, T., Nicol, C., Rothschild, M., and Moore, B.P. The Biological Roles of Pyrazines: Evidence for a Warning Odour Function. *Biological Journal of the Linnean Society* **1987**, *31*, 113–28.

Guo, D.; Zhang, D.; Li, N.; Zhang, L.; Yang, J. A Novel Breath Analysis System Based on Electronic Olfaction. *IEEE Transactions on Biomedical Engineering* **2010**, *57* (11), 2753–2763

Haaretz Service. Israel Develops 'Skunk Bomb' for Riot Control Situations. Available: [http://www.haaretz.com/news/israel-develops-skunk-bomb-for-riot-control-situations-1.135003]. Confirmed on June 28, 2012.

Hammond, J. Introducing the Wine Aroma Wheel. Available: [http://www.examiner.com/article/introducing- the-wine-aroma-wheel]. Confirmed on Aug. 21, 2012.

Haswell, J. *The Intelligence and Deception of the D-Day Landings*; London, UK: Batsford Publishers, 1979.

Haswell, J. *The Tangled Web: The Art of Tactical and Strategic Deception;* Wendover, UK: John Goodchild Publishers, 1985.

Helm, B. Finding Land Mines by Following Bees. Bloomberg Businessweek. Available: [http://www.businessweek.com/stories/2005-08-15/finding-land-mines-by-following-a-bee.htm]. Confirmed Aug. 9, 2012.

Henning, H. *Der Geruch*. Leipzig, Germany: Barth, 1916.

Herz, R.S. Sex Differences in Response to Physical and Social Factors Involved in Human Mate Selection: The Importance of Smell for Women. *Evolution and Human Behavior* **2002**, *23* (5), 359–364.

Herz, R. S. A Naturalistic Analysis of Autobiographical Memories Triggered by Olfactory, Visual, and Auditory Stimuli. *Chemical Senses* **2004**, *29* (3), 217–224.

Herz, R. S. *The Scent of Desire*; New York, NY: Harper-Collins Publishers, 2008.

Hoballah, M. E.; Turlings, T.C.J. The Role of Fresh Versus Old Leaf Damage in the Attraction of Parasitic Wasps to Herbivore-induced Maize Volatiles. *Journal of Chemical Ecology* **2005**, *31* (9), 2003–2018.

Hoffman, H. J.; Ishii, E. K.; MacTurk, R. H. Age-related Changes in the Prevalence of Smell/Taste Problems Among the United States Adult Population: Results of the 1994 Disability Supplement to the National Health Interview Survey (NHIS). *Annals of the New York Academy of Science* **1998**, *855*, 716–722.

Hughes, N. K.; Price, C. J.; Banks, P. B. Predators are Attracted to the Olfactory Signals of Prey. *PLoS ONE* **2010**, *5* (9), e13114.

Jolly, N. P.; Hattingh, S. A Brandy Aroma Wheel for South African Brandy. *South African Journal of Enology and Viticulture* **2001**, *22* (1), 16–21.

Jones, M. B.; Fjellanger, R.; Cox, C.; Poling, A. Remote Explosive Scent Tracing of Explosive Remnants of War: A Perspective from 2010 Morogoro Workshop. *Journal of ERW and Mine Action*, 15.1. Available: [http://www.hdic.jmu.edu/journal/15.1/r_d/jones/jones.htm]. Confirmed on July 28, 2012.

Jürgens, A.; El-Sayed, A. M.; Suckling, M. Do Carnivorous Plants use Volatiles for Attracting Prey Insects? *Functional Ecology* **2009**, *23* (5), 875–887.

Karstensen, H.; Tommerup, N. Isolated and Syndromic Forms of Congenital Anosmia. *Clinical Genetics* **2001**, *81* (3), 210–215.

Kats, L. B.; Dill, L. M. The Scent of Death: Chemosensory Assessment of Predation Risk by Prey Animals. *Ecoscience* **1998**, *5* (3), 361–394.

Kaitz, M.; Good, A.; Rokem, A. M.; Eidelman, A. Mothers' Recognition of Their Newborns by Olfactory Cues. *Developmental Psychobiology* **1987**, *20* (6), 587–591.

Kieran, M. The Vice of Snobbery: Aesthetic Knowledge Justification and Virtue in Art Appreciation. *Philosophical Quarterly* **2010**, *60* (239), 243–263.

Kim, H.; Park, J.; Noh, K.; Gardner, C. J.; Kong, S. D.; Kim, J.; Jin, S. X–Y Addressable Matrix Odor-releasing System Using an On–Off Switchable Device. *Angewandte Chemie (International Edition)* **2011**, *50*, 6771–6775.

Kirby, R. Operation Snoopy: The Chemical Corps' "People Sniffer". *Army Chemical Review* **2007**, *1*, 20–22.

Knols, B.G.J.; Takken, W.; Cork, A.; Jong R. D. Odour Mediated, Host Seeking Behaviour of *Anopheles* Mosquitoes: A New Approach. *Annals of Tropical Medicine and Parasitology* **1997**, *91* (Supl 1), 117–118.

Krall, B. S.; Bartelt, R. J.; Lewis, C. J.; Whitman, D. W. Chemical Defense in the Stink Bug *Cosmopepla bimaculata*. *Journal of Chemical Ecology* **1999**, *25* (11), 2477–2494.

Krusemark, E. A.; Li, W. Enhanced Olfactory Sensory Perception of Threat in Anxiety: An event-related fMRI study. *Chemosensory Perception* **2012**, *5*, 37-45.

Laino, C. Dogs Sniff Out Prostate Cancer. Available: [http://www.webmd.com/prostate-cancer/news/20100602/dogs-sniff-out-prostate-cancer]. WebMD Health News, 2012.

Larsson, M.; Willander. J. Autobiographical Odor Memory. *Annals of the New York Academy of Sciences* **2009**, *1170*, 318–323.

Latimer, J. *Deception in War: The Art of the Bluff, the Value of Deceit, and the Most Thrilling Episodes of Cunning in Military History, from the Trojan Horse to the Gulf War*; New York, NY: The Overlook Press, 2003.

Laumbach, R. J.; Kipen, H. M.; Kelly-McNeil, K.; Zhang, J.; Zhang, L.; Lioy, P. J.; Ohman-Strickland, P.; Gong, J.; Kusnecov, A.; Fiedler, H. Sickness Response Symptoms Among Healthy Volunteers After Controlled Exposures to Diesel Exhaust and Psychological Stress. *Environmental Health Perspectives* **2011**, *119* (7), 945–950.

Layton, J. How Search and Rescue Dogs Work. Available: [http://science.howstuffworks.com/environmental/life/zoology/mammals/sar-dog1.htm]. Confirmed on August 10, 2012.

Lenochová, P.; Vohnoutová, P.; Roberts, S. C.; Oberzaucher, E.; Grammer, K.; Havlíček, J. Psychology of Fragrance Use: Perception of Individual Odor and Perfume Blends Reveals a Mechanism for Idiosyncratic Effects on Fragrance Choice. *PLoS ONE* **2012**, *7* (3), e33810.

Letowski, T. Owning the Environment: Stealth Soldier—Research Outline. ARL Special Report ARL-SR-242 (May 2012). Aberdeen Proving Ground, MD: ARL.

Li, W.; Moallem, I.; Paller, K. A.; Gottfried, J. A. Subliminal Smells Can Guide Social Preferences. *Psychological Science* **2007**, *18* (12), 1044–1049.

Liljenquist, K.; Zhong, C. B.; Galinsky A. D. The Smell of Virtue: Clean Scents Promote Reciprocity and Charity. *Psychological Science* **2010**, *21* (3), 381–383.

Lindstrom, L.; Rowe, C.; Guilford, T. Pyrazine Odour Makes Visually Conspicuous Prey Aversive. *Proceedings of the Royal Society of London Series B* **2001**, *268*, 159–162.

Lucretius (58BC/2010). *De Rerum Natura* [*On the Nature of Things*]. English translation by Ian Johnston. Alington, VA: Richer Resources Publications.

Mahmut, M. K.; Stevenson, R. J. Olfactory Abilities and Psychopathy: Higher Psychopathy Scores are Associated with Poorer Odor Discrimination and Odor Identification. *Chemosensory Perception.* DOI: 10.1007/s12078-012-9135-7, 2012.

Mateo, J. M. Kin Recognition in Ground Squirrels and Other Rodents. *Journal of Mammalogy* **2003**, *84* (4), 1163–1181.

McManners, H. Israelis Invent Stink Bomb for Riot Control. *The Independent* (18 September 2004). Available: [http://www.independent.co.uk/news/world/middle-east/israelis-invent-stink-bomb-for-riot-control-546665.html]. Confirmed on July 20, 2012.

Mehlis, M.; Bakker, T.C.M.; Frommen, J. G. Smells Like Sib Spirit: Kin Recognition in Three-spined Sticklebacks (*Gasterosteus aculeatus*) is Mediated by Olfactory Cues. *Animal Cognition* **2008**, *11* (4), 643–650.

Meilgaard, M. C.; Reid, D. S.; Wyborski, K. A. Reference Standards for Beer Flavor Terminology System. *Journal of the American Society of Brewing Chemists* **1982**, *40*, 119–128.

Mihm, S. The Year in Ideas: Stench Warfare. *The New York Times* (15 December 2002, p. E65), **2002**.

Milinski, M.; Wedekind, C. Evidence for MHC-correlated Perfume Preferences in Humans. *Behavioral Ecology* **2001**, *12* (2), 140–149.

Mohamed, E. I.; Linder, R.; Perriello, G.; Di Daniele, N.; Pöppl, S. J.; De Lorenzo, A. Predicting Type 2 Diabetes Using an Electronic Nose-based Artificial Neural Network Analysis. *Diabetes, Nutrition, and Metabolism* **2002**, *15* (4), 215-221.

Moncrieff, R. W. What is Odor? A New Theory. *American Perfumer* **1949a**, *54*, 453.

Moncrieff, R. W. A New Theory of Odour. *Perfumery and Essential Oil Record* **1949b**, *40*, 279–285.

Moore, C. H. Zinc Nanoparticle Enhancement of the Olfactory Neuron Response to Odorants Associated with Explosives. Master's thesis. Department of Anatomy, Physiology, and Pharmacology. Auburn University, AL, 2011.

Moore, B. P.; Brown, W. V.; Rothschild, M. Methylalkylpyrazines in Aposematic Insects, Their Hostplants and Mimics. *Chemoecology* **1990**, *1*, 43–51.

Morgan, E.D. Trail Pheromones of Ants. *Physiological Entomology* **2009**, *34* (1), 1–17.

Morran, C. Samsung and Science Work Together to Make Smell-O-Vision a Reality. *The Consumerist*. Available: [http://consumerist.com/2011/06/samsung-and-science-work-together-to-make-smell-o-vision-a-reality.html]. Confirmed on July 28, 2012.

Muir, D. D.; Hunter, E. A.; Williams, S.A.R.; Brennan, R. N. Sensory Profiles of Commercial Fruit Juice Drinks: Influence of Sweetener Type. *Journal of Science of Food and Agriculture* **1998**, *77*, 559–565.

Mujica-Parodi, L. R.; Strey, H. H.; Frederick, B.; Savoy, R.; Cox. D.; Botanov, Y.; Tolkunov, D.; Rubin, D.; Weber, J. Chemosensory Cues to Conspecific Emotional Stress Activate Amygdala in Humans. *PLoS ONE* **2009**, *4* (7), 36415.

Murphy, C. Taste and Smell in the Elderly. In: H.L. Meiselman and R.S. Rivlen (eds.), *Clinical Measurement of Taste and Smell*, pp. 343–371. New York, NY: MacMillin, 1983.

Müller-Schwarze, D.; Altieri, R.; Porter, N. Alert Odor from Skin Gland in Deer. *Journal of Chemical Ecology* **1984**, *10* (12), 1707–1729.

NHDES. *Odors and Your Health*. Environmental Fact Sheet ARD-EHP-24. Concord, NH: New Hampshire Department of Environmental Services, 2012.

Noble, A. C.; Arnold, R. A.; Masuda, B. M.; Pecore, S. D.; Schmidt, J. O.; Stern, P. M. Progress Towards a Standardized System of Wine Aroma Terminology. *American Journal of Enology and Viticulture* **1984**, 35, 107–109.

Nolte, D. L.; Mason, J. R.; Epple, G.; Aronov, E.; Campbell, D. L. Why are Predator Urines Aversive to Prey? *Journal of Chemical Ecology* **1994**, *20* (7), 1505–1516.

Oakes, H. International K-9 Search-and-Rescue Services for Missing People and Pets. IK9SARS. Available: [http://www.k9sardog.com/faqs.html]. Confirmed on August 10, 2012.

Orfici, D. *A Guide to Mine Action*. Geneva (Switzerland): Geneva International Centre for Human Demining (GICHD), 2003.

Pain, S. Stench Warfare. *New Scientist* **2001**, *171* (2298), 42–45 (7 July 2001).

Papes, F.; Logan, D. W.; Stowers, L. The Vomeronasal Organ Mediates Interspecies Defensive Behaviors Through Detection of Protein Pheromone Homologs. *Cell* **2010**, *141*, 692–703.

Parejo, D.; Amo, L.; Rodríguez, J.; Avilés, J. M. Rollers Smell the Fear of Nestlings. *Biology Letters* **2012**, *8* (4), 502–504.

Piggot, J. R.; Jardine, S. P. Descriptive Sensory Analysis of Whiskey Flavor. *Journal of the Institute of Brewing* **1979**, *85*, 82–85.

Pauling, L. *The Nature of the Chemical Bond and the Structure of Molecules and Crystals;* New York, NY: Cornell University Press, 1939.

Peschke, K. Male Aggression, Female Mimicry and Female Choice in the Rove Beetle *Aleochara curtula* (Coleoptera, Staphylinidae). *Ethology* **1987**, *75* (4), 265–284.

Rheingold, H. *Virtual Reality*; New York, NY: Simon and Schuster, 1992.

Philpott, C. M.; Bennett, A.; Murty, G. E. A Brief History of Olfaction and Olfactometry. *The Journal of Laryngology and Otology* **2008**, *134* (7), 657–662.

Rossiter, K .J. Structure-odor Relationships. *Chemical Reviews* **1996**, *96* (8), 3201–3240.

Rosenfeld, P. E.; Clark, J. J.; Hensley, A. R.; Suffet, I. H. The Use of an Odour Wheel Classification for the Evaluation of Human Health Risk Criteria for Compost Facilities. *Water Science Technology* **2007**, *55* (5), 345–357.

Rosture, J. Aboriginals - The Body as Living Art. Available: [http://www.janesoceania.com/australia_aboriginal_bodylivingart/index1.htm]. Confirmed on June 24, 2012.

Rubin, D.; Botanov, Y.; Hajcak, G.; Mugica-Parodi, L. R. Second-hand Stress: Inhalation of Stress Sweat Enhances Neural Response to Neutral Faces. *Social Cognitive and Affective Neuroscience* **2012**, *7* (2), 208–212.

Salthammer, T.; Bahadir, M. Occurrence, Dynamics and Reactions of Organic Pollutants in the Indoor Environment. *CLEAN – Soil, Air, Water* **2009**, *37*, 417–435.

Santos, P. S.; Schinemann, J. A.; Gabardo, J.; Bicalho Mda, G. New Evidence that the MHC Influences Odor Perception in Humans: A Study with 58 Southern Brazilian Students. *Hormones and Behavior* **2005**, *47* (4), 384–388.

Scentcom Homepage. http://www.scentcom.co.il (accessed Dec 2012).

Schiestl, F. P.; Peakall, R.; Mant, J. G.; Ibarra, F.; Schulz, C.; Franke, S.; Francke, W. The Chemistry of Sexual Deception in an Orchid-Wasp Pollination System. *Science* **2003**, *302*, 437–438.

Schönrogge, K.; Wardlaw, J. C.; Peters, A. J.; Everett, S.; Thomas, J. A.; Elmes, G. W. Changes in Chemical Signature and Host Specificity from Larval Retrieval to Full Social Integration in the Myrmecophilous Butterfly *Maculinea rebeli*. *Journal of Chemical Ecology* **2004**, *30* (1), 91–107.

Sense of Smell Institute. Welcome to the Sense of Smell Institute. Available: [http://www.sensofsmell.org]. Confirmed on August 3, 2012.

Shepherd, G. M. The Human Sense of Smell: Are we better than we think? *PLoS Biology* **2004**, *2* (5), e146.

Simons, R.R.; Felgenhauer, B. E.; Jaeger, R. G. Salamander Scent Marks: Site of Production and Their Role in Territorial Defence. *Animal Behaviour* **1994**, *48* (1), 97–103.

Smith, R.J.F. Alarm Signals in Fishes. *Reviews in Fish Biology and Fisheries* **1992**, *2*, 33–63.

Sorensen, J. S.; Heward, E.; Dearing, M. D. Plant Secondary Metabolites Alter the Feeding Patterns of a Mammalian Herbivore (*Neotoma lepida*). *Oecologia* **2005**, *146* (3), 415–422.

Spengler, J. D.; McCarthy, J. F.; Samet, J. M. *Indoor Air Quality Handbook*; New York, NY: McGraw-Hill, 2000.

Stoddart, D. M. *The Scented Ape: The Biology and Culture of Human Odour*; Cambridge, UK: Cambridge University Press, 1990.

Suffet, I. H.; Rosenfeld, P. The Anatomy of Odour Wheels for Odours of Drinking Water, Wastewater, Compost and the Urban Environment. *Water Science and Technology* **2007**, *55* (5), 335–344.

Teixeira, M. A.; Rodríguez, O.; Rodrigues, A. E. Perfumery Radar: A Predictive Tool for Perfume Family Classifications. *Industrial and Engineering Chemistry Research* **2010**, *49* (22), 11764–11777

Tchoukanov, A. Smell – Search and Rescue. Available: [http://www.tchoukanov.com/?page_id=20]. Confirmed on July 28, 2012.

Trivedi, B.P. U.S. Military is Seeking Ultimate "Stink Bomb". National Geographic Today (Jan 7, 2002). Available: [http://news.nationalgeographic.com/news/2002/01/0107_020107TVstinkbomb.html]. Confirmed on June 28, 2012.

Turin, L. A Spectroscopic Mechanism for Primary Olfactory Reception. *Chemical Senses* **1996**, *21* (6), 773-791.

Turin, L. What You Can't Smell Will Kill You; *The New York Times*, 21 January 2007, p. C13, 2007.

Vokshoor, A.; McGregor, J. Olfactory System Anatomy. Available: [http://emedicine.medscape.com/article/835585-overview]. Confirmed on June 28, 2012.

Wang, F.; Nemes, A.; Mendelsohn, M.; Axel, R. Odorant Receptors Govern the Formation of a Precise Topographic Map. *Cell* **1998**, 93, 47–60.

Wanzala, W.; Sika, N.F.K.; Gule, S.; Hassanali, A. Attractive and Repellent Host Odours Guide Ticks to Their Respective Feeding Sites. *Chemoecology* **2004**, *14*, 229–23.

Wedekind, C.; Füri, S. Body Odour Preferences in Men and Women: Do They Aim for Specific MHC Combinations or Simply Heterozygosity? *Proceedings of the Royal Society of London Series B* **1997**, *264*, 1471–1479.

White, A. M.; Swaisgood R. R.; Zhang, H. The Highs and Lows of Chemical Communication in Giant Pandas (*Ailuropoda melanoleuca*): Effect of Scent Deposition Height on Signal Discrimination. *Behavioral Ecology and Sociobiology* **2002**, *51*, 519–52.

Williams, A. A. The Development of a Vocabulary and Profile Assessment Method for Evaluating the Flavor Contribution of Cider and Perry Aroma Constituents. *Journal of Science of Food and Agriculture* **1975**, *26* (5), 567–582.

Wise, P. M.; Olsson, M. J.; Cain, W. S. Quantification of Odor Quality. *Chemical Senses* **2000**, *25* (4), 429–443.

Wisenden, B. D. Olfactory Assessment of Predation Risk in the Aquatic Environment. *Philosophical Transactions of the Royal Society of London Series B* **2000**, *255*, 1205–1208.

Wolfe, J. M.; Kluender, K. R.; Levi, D. M. Bartoshuk, L. M.; Herz, R. S.; Klatzky, R. L.; Lederman, S. J.; and Merfeld, D. M. *Sensation and Perception* (3rd ed.); Sunderland, MA: Sinauer Associates, 2011.

Yanagida, Y.; Kawato, S.; Nona, H.; Tomono, S. A.; Tetsutani, N. Projection-based Olfactory Display with Nose Tracking. *Proceedings of the 2004 IEEE Virtual Reality Conference*, pp, 43-50. Chicago, IL: IEEE, 2004.

Zala, S. M.; Potts, W. K.; Penn, D. J. Scent-marking Displays Provide Honest Signals of Health and Infection. *Behavioral Ecology* **2004**, *15* (2), 338–344.

Zald, D. H. The Human Amygdala and the Emotional Evaluation of Sensory Stimuli. *Brain Research Review* **2003**, *41*, 88–123.

Zald, D.;H.; Pardo, J.V. Emotion, Olfaction, and the Human Amygdala: Amygdala Activation During Aversive Olfactory Stimulation. *Proceedings of the National Academy of Sciences* **1997**, *94*, 4119–4124.

Zhang, Y. H.; Zhang, J. X. Urine-derived Key Volatiles May Signal Genetic Relatedness in Male Rats. *Chemical Senses* **2011**, *36* (2), 125–135.

Zuri, I.; Gazit, I.; Terkel, J. Effect of Scent-marking in Delaying Territorial Invasion in the Blind Mole-rat. *Spalax ehrenbergi. Behaviour* **1997**, *134* (11–12), 867–880.

Zwaardemaker, H. *Die Physiologie de Geruchs*; Leipzig, Germany: Engelmann, 1895.

Zwaardemaker, H. *L'Odorat*; Paris, France: Doin, 1925.

List of Symbols, Abbreviations, and Acronyms

ANN	artificial neural networks
APOPO	Anti-Persoonsmijnen Ontmijnende Product Ontwikkeling
ARL	U.S. Army Research Laboratory
ASTM	American Society of Testing and Materials
ASR	active smell reduction
CBP	Customs and Border Protection
DARPA	Defense Advanced Research Projects Agency
e-SAR	electronic search and rescue
IBIS	Identification Based on Individual Scent
ICT	Institute for Creative Technologies
MHC	major histocompatibility complex
MISL	Media Information Science Laboratories
OPSEC	operations security
QMB	quartz microbalance
REST	Remote Explosive Scent Tracing
SAR	search and rescue
UCSD	University of California, San Diego
UXO	unexploded ordinance
VR	virtual reality